Geochemical Evolution of Guatemala's Ophiolitic Belts

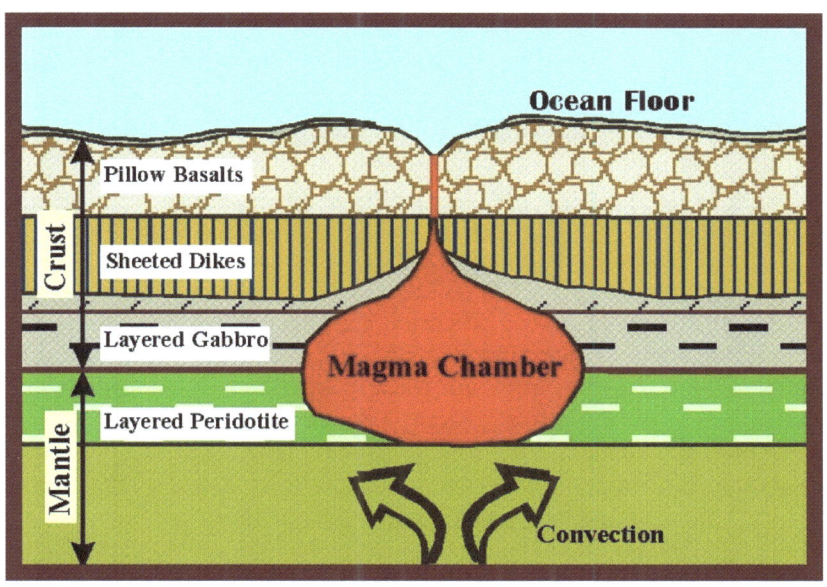

Ricardo A. Valls, P. Geo., M. Sc.

Mente et Maleo... and Computers.

Editing:	Liudmila V. Valls
Design and layout:	Ricardo A. Valls
Photos and illustrations:	Ricardo A. Valls, Hugo Matzer, Julio Roberto Perez (RIP)

Cover: Schematic model of the formation of an ophiolitic belt.

Copyright © 2016 Ricardo A Valls

All rights reserved. This publication is protected by copyright and permission should be obtained from Valls Geoconsultant prior to any prohibited reproduction, storage in a retrieval system, or transmission in any form or by any means, electronic, mechanical, photocopying, recording, or likewise. For information regarding permission write to Valls Geoconsultant at 1008-299 Glenlake Ave., Toronto, Ontario, Canada, E-mail: vallsvg@gmail.com.

ISBN-13: 978-1534717220
ISBN-10: 1534717226

Table of Contents

Introduction 4
Ophiolitic Belts in Guatemala 6
    Petrologic Classifications of Ophiolites 9
    Ophiolitic Phases 12
    Ophiolitic Ages 17
    Ophiolites Around the World 19
Comparing Guatemalan Ophiolitic Belts 20
    The North and South Motagua Ophiolitic Complex 20
    The Juan de Paz – Los Mariscos Ophiolitic Complex 21
    The Sierra de Santa Cruz Ophiolitic Complex 23
Products of Serpentinization 25
Petrography of the Guatemalan Ophiolitic Belts 28
    Olivine-rich Lherzolite 28
    Lherzolite 29
    Olivine-rich Websterite 30
    Hazburgite 31
Model for the Evolution of the Guatemalan Ophiolitic Belts 34
    Phase I- Magmatic Chamber 34
    Phase II- Obduction 34
    Phase III- Serpentinization Process 35
    Phase IV- Oxidation process 36
Unique Characteristics of the Geosol Izabal 38
References 43
Notes 44
About the Author 45

Introduction

The aim of the present paper is to provide an overview of the ophiolitic processes associated with the Motagua Suture Zone in Guatemala and to show their nickel lateritic potential. It is intended to be an accompanied document to the Field Trip Guidebook (6th Edition).

In preparing this paper I have compiled information from different sources, including several internet sources, combined with my personal experiences in the mapping of the area and help and cooperation of the geologists from the Guatemalan Ministry of Energy and Mines.

The Caribbean Plate (Fig. 1) is the result of the Mesozoic-Present interaction of the Nazca, Cocos, North, and South American plates. The margins of these plates are represented by large deformed belts which resulted from several compressional episodes that started in the Cretaceous and were followed by tensional and strike-slip tectonics.

The present day north-western margin of the Caribbean Plate crops out in Guatemala along the Motagua Suture Zone (MSZ). This zone links the Meso-American trench with the Cayman Islands extensional system as shown in Fig. 2.

The MSZ represents a sinistral shear-zone between the Maya Continental Block (MCB) to the north and the Chortis Continental Block (ChCB) to the south. The MSZ includes the Motagua Fault Systems of Polochic, Motagua, Cabañas, and Jocotán. All these are E-W and ENE-WSW strike-slip faults. Some of them are still seismically active. The MSZ also includes E-W uplift structures (Sierra de Chuacús, Sierra de Las Minas, and Montañas del Mico), pull-apart basins like the one responsible for the formation of the Lake Izabal, and N-S oriented grabens (Chiquimula, Guatemala, etc.).

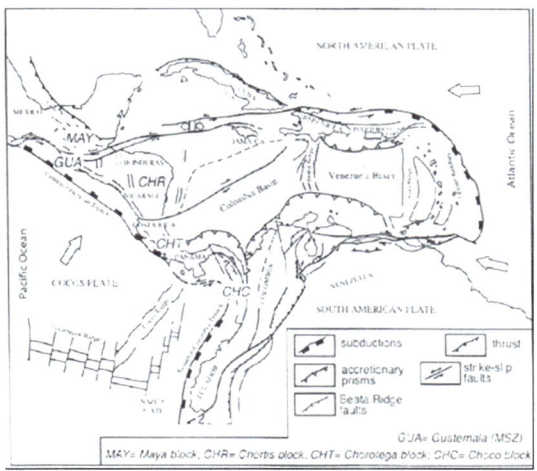

Figure 1. Structural sketch map of the Caribbean area (from Giunta et al., 2002).

Figure 2. Bathometric representation of the Caribbean Plate and its relationship with the neighboring plates and structures.

Ophiolitic Belts in Guatemala

An ophiolite is a thrust sheet of ancient oceanic lithosphere which has been obducted over the continental crust in the course of orogeny. The majority of geologists interpret these sequences as oceanic crustal and upper mantle material that has been pushed up onto continents when slivers of the sea floor were caught between converging plates (Fig. 3). The term "ophiolite" comes from the Greek word "snake stone".

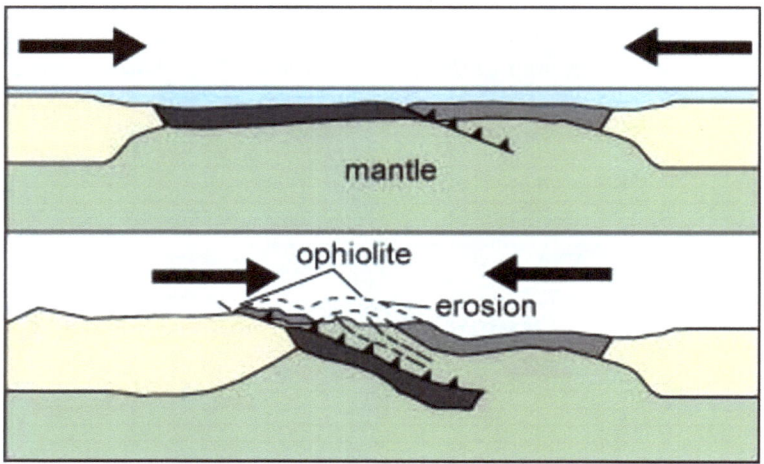

Figure 3. Diagram showing the model of formation of an ophiolitic belt. Photo courtesy of
http://volcano.und.nodak.edu/vwdocs/vw_hyperexchange/ophiolites.htm.

An ophiolite is a stratified igneous rock complex composed of upper basalt member, middle gabbro member, and lower peridotite member. Some large complexes measure over 10 km thick, 100 km wide, and 500 km long. Basalt and gabbro are commonly altered into patchy green rocks, and peridotite is mostly changed into black, greasy serpentinite.

The ophiolite succession can be correlated with the seismologic layering of the oceanic lithosphere (Fig. 4). The sedimentary cover corresponds to Layer 1, basaltic pillow lava matches Layer 2, sheeted dikes and gabbro with occasional plagiogranite intrusions are correlated to Layer 3, and ultramafic cumulates and residual mantle peridotite represent Layer 4 (mantle).

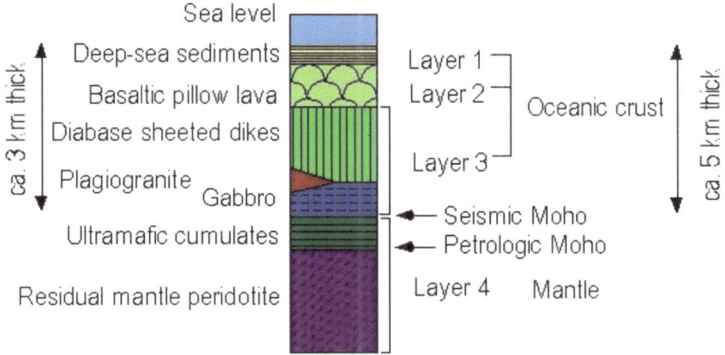

Figure 4. Typical ophiolitic succession.

Ophiolites usually occur as a nappe (intact thrust sheet) or as a mélange (tectonic mixture of fragments). In collisional orogenic belts, ophiolites generally lie on older continental basement, while in circum-Pacific orogenic belts ophiolites generally lie on younger accretionary complexes. In Guatemala, they occur as mélanges over older rocks (Fig. 5).

Following is a list of the most common minerals related to the ophiolitic processes. The list is by no means exhaustive and serves only to explain the principal constituents of the ultramafic rocks which may be involved in the process of serpentinization, or may be present as accessory minerals.

Albite	$NaAlSi3O_4$
Anorthite	$CaAl_2Si_2O_4$
Augite	$(Ca, Mg, Fe, Al)_2(Al, Si)_2O_6$
Awaruite	Ni_3Fe

Bastite	$Mg_2Si_2O_6$
Brucite	$Mg(OH)_2$
Chromite	$FeCr_2O_4$
Clinopyroxene	$Ca(Mg, Fe)Si_2O_6$
Diopside	$(Ca, Mg)Si_2O_6$
Fayalite	$FeSiO_4$
Fe-Chrysotile	$Fe_3Si_2O_5(OH)_4$
Forsterite	$MgSiO_4$
Garnerite	$H_2(Ni,Mg)SiO_4$
Hyperstene	$(Mg, Fe)SiO_3$
Limonite	$[(Fe, Ni)O(OH)]$
Magnetite	Fe_3O_4
Milerite	NiS
Orthopyroxene	$(Mg, Fe)SiO_3$
Pentlandite	$[(Ni,Fe)_9S_8]$
Serpentinite	$Mg_2Si_2O_5(OH)_4$
Siderite	$FeCO_3$
Talc	$MgSi_4O_5(OH)_4$

Figure 5. Ophiolitic belts in Guatemala.

Petrologic Classifications of Ophiolites

Ophiolites may have formed either at divergent plate boundaries (mid-oceanic ridges) or at convergent plate boundaries (supra-subduction zones; i.e. island arcs and marginal basins). They are called MOR and SSZ types, respectively. These types are identified by chemical composition of the rocks and minerals in comparison with those from various tectonic settings on the earth at present.

9

Ophiolitic mantle peridotite is the refractory residue after extraction of basaltic melt through partial melting processes in the mantle. Although primary mantle peridotite may be lherzolite with abundant clinopyroxene, it changes into clinopyroxene-poor (or - free) harzburgite as the degree of melting increases (Fig. 6). The mantle peridotite samples dredged from the mid-oceanic ridges are mostly lherzolite, while those dredged from supra-subduction zones (trench walls) are mostly harzburgite.

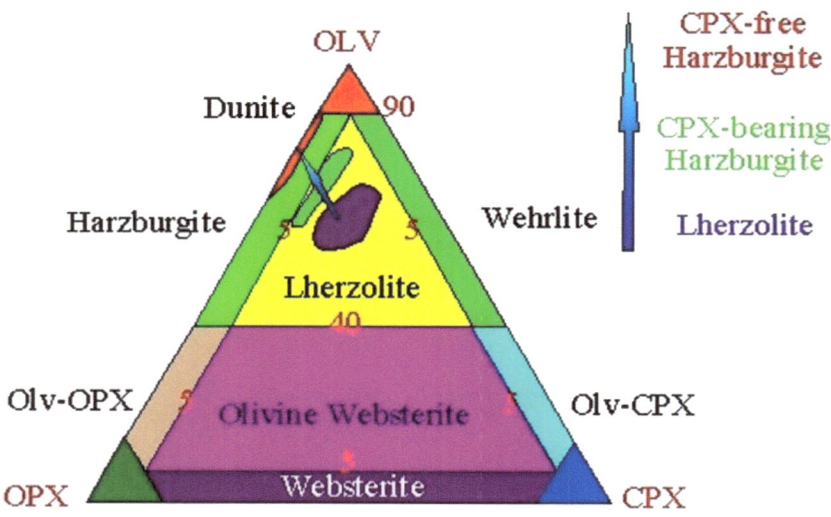

Figure 6. Modal variation of residual mantle peridotite with increasing degree of melting.

In Guatemala, only the South and North Motagua ophiolitic belts are classified as harzburgites (trench wall), while the rest are layered intrusives alternating olivine-rich lherzolites; normal lherzolites and websterites more typical of an Island Arc environment.

Ophiolitic igneous cumulates show systematic variation in the crystallization sequence of minerals corresponding to the petrologic diversity of the underlying peridotite mantle. The mineral crystallizing next to olivine typically varies from plagioclase through clinopyroxene to orthopyroxene as the degree of melting in the underlying mantle increases (Fig. 6). The characteristic cumulate rocks correspondingly vary from wehrlite to harzburgite.

In general, ophiolitic basalt also varies from alkali basalt or high-alumina basalt (like mid-ocean ridge basalt (MORB)) through low-alumina basalt (like island-arc tholeiite (IAT)) to boninite (high-magnesia andesite) in correspondence with the petrologic variation of the underlying members (Fig. 7).

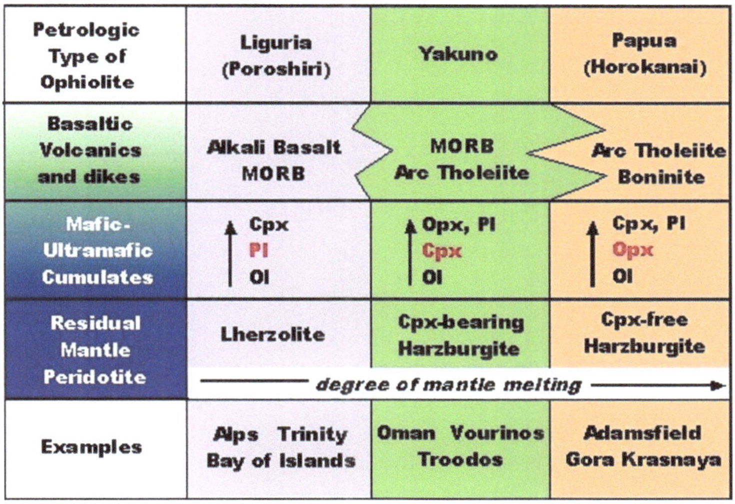

Petrologic Type of Ophiolite	Liguria (Poroshiri)	Yakuno	Papua (Horokanai)
Basaltic Volcanics and dikes	Alkali Basalt MORB	MORB Arc Tholeiite	Arc Tholeiite Boninite
Mafic-Ultramafic Cumulates	Cpx / Pl / Ol	Opx, Pl / Cpx / Ol	Cpx, Pl / Opx / Ol
Residual Mantle Peridotite	Lherzolite	Cpx-bearing Harzburgite	Cpx-free Harzburgite
	← degree of mantle melting →		
Examples	Alps Trinity Bay of Islands	Oman Vourinos Troodos	Adamsfield Gora Krasnaya

Figure 7. Systematic variations of the ophiolitic igneous cumulates.

The Guatemalan ophiolitic belts are characterized by extreme petrologic diversity. Juxtaposition of olivine- depleted websterites and olivine-rich Lherzolites is common for the youngest belts.

Ophiolitic Phases

The following petrologic assemblages are usually associated with an ophiolitic profile (Fig. 8).

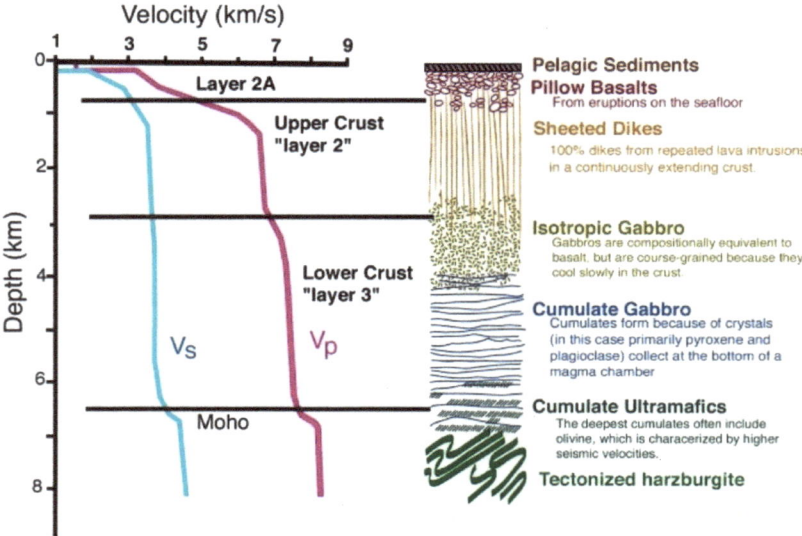

Figure 8. Representation of the ocean crust. The seismic structure (from seismic experiments) is combined with the lithology of ophiolites.

The Pillow Basalt Layer (PBL) is well represented in some ophiolitic belts in Guatemala, while it is absent (or have not yet been found) in others. The best example of the PLB can be found associated to the Huehuetenango ophiolitic belt (Fig. 9). Pseudo pillow lavas can be also found associated to the North Motagua ophiolitic belt (Fig. 10).

Figure 9. Outcrop of pillow basalts of the Huehuetenango ophiolitic belt.

Figure 10. Pseudo pillow lavas from the North Motagua ophiolitic belt.

According to the established models (Fig. 11), we can find VMS type of deposits associated to the PBL. The Sierra de Santa Cruz ophiolitic belt hosts such type of deposit, the Cyprus type copper deposit of Oxec.

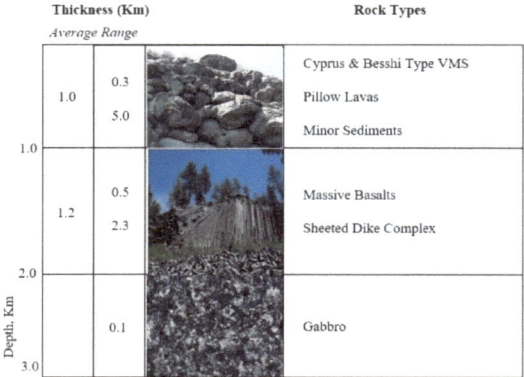

Figure 11. Schematic ophiolitic complex.

The Sheeted Dike Layer (SDL) is best represented again at the Huehuetenango ophiolitic belt (Fig. 12). The unit consists of 100% dikes with no intervening screens of other rocks. Possibly related to the underlying gabbros, they are not derived directly from them.

Figure 12. The sheeted dike layer of the Huehuetenango ophiolitic belt.

The Cumulate Gabbro Layers (CGL) are generally mafic at the base and grade upward to more feldspathic rocks. Olivine becomes progressively more iron-rich, reaching Fo70 in the uppermost gabbros. Orthopyroxene is present but far less abundant than in the underlying deformed ultramafic rocks. Diopside augite is the principal pyroxene. Toward the top of the layered gabbros, irregular intrusive bodies of plagiogranites occur. These are thought to be the final differentiation product of the gabbroic magma, which gave rise to layered rocks. In Guatemala we have found cumulate gabbros only near the River Cahabón within the Sierra de Santa Cruz ophiolitic belt.

The Ultramafic Cumulate Layers (UCL) are usually located below the lowest layered mafic, the transition to the harzburgitic mantle section, and are usually marked by the occurrence of plastically deformed dunites. The deformation, which is characteristic of the underlying harzburgite, dies out in this transition zone. In terms of mineral composition, the mafic-ultramafic transition is smoothed by the presence of ultramafic layers, gabbro dikes, sills and impregnations in the dunites and the uppermost harzburgites. The amount of clinopyroxene and plagioclase increase upward through the zone. Thus, dunite (olv) at the base passes up through wherlite (olv+cpx) to clinopyroxenite and troctolite (plag+olv). With the decrease in the amount of olivine, then, the degree of deformation decreases.

Figure 13. Plastically deformed harzburgites from the North Motagua ophiolitic belt.

The Metamorphic Envelope (ME). These are dynamo metamorphic rocks welded to the base of ophiolites (metamorphic soles or dynamo thermal aureoles). They are generally believed to have formed during detachment and/or emplacement of oceanic lithosphere into orogenic belts. The rocks found here range from mylonitic peridotites to garnet amphibolites, epidote amphibolites, and greenschists. In Guatemala we find greenschist facies associated to all ophiolitic belts (Figs. 14-15), garnet amphibolites associated to the North Motagua ophiolitic belt (Fig 16), and the best example of mylonitization can be found at the Juan de Paz – Los Mariscos ophiolitic belt (Fig. 17).

Figure 14. Greenschist facies at Huehuetenango ophiolitic belt.

Figure 15. Greenschist facies of the South Motagua ophiolitic belt.

Figure 16. Garnet amphibolites associated with the North Motagua ophiolitic belt.

Figure 17. Zone of intense mylonitization at the Juan de Paz - Los Mariscos ophiolitic belt.

Ophiolitic Ages

Reported formation ages of ophiolites show three distinct peaks at about 750, 450 and 150 Ma, respectively (Fig. 18). These are called ophiolite pulses. Each pulse corresponds to the period of worldwide magmatic event as represented by voluminous granite intrusions. Production rate of oceanic crust was distinctly high during the 80 and 120 Ma interval of Cretaceous time, as evidenced by wide area of the ocean floor formed in this interval. This interval corresponds to the latter half of the Mesozoic ophiolite pulse.

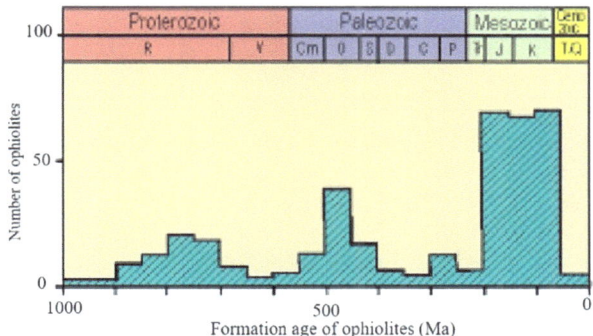

Figure 18. Ophiolitic "pulses" in Earth's history.

Age determinations from micas from the ultramafic rocks from the North Motagua and the Juan de Paz – Los Mariscos ophiolitic complexes have shown ages varying in a west-east direction towards younger rocks (Fig. 19). The same trend is observed in Cuba. Following the same logic, I expect the Huehuetenango ophiolitic complex to be older than the Baja Verapaz complex, and in turn older than the Sierra de Santa Cruz ophiolitic complex at the eastern end of the Polochic Fault.

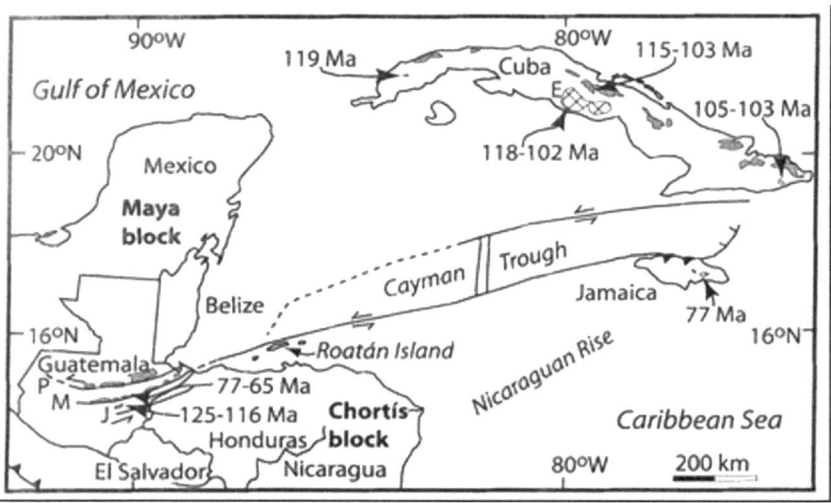

Figure 19 Age determinations of the ophiolites within the Caribbean Plate according to Harlow, E.G. et al., 2004.

Ophiolites Around the World

Ophiolites issued by each pulse tend to form a particular ophiolite belt. Late Proterozoic (ca. 750 Ma) ophiolites are distributed in the Pan-African orogenic belt, early Paleozoic (ca. 450 Ma) ophiolites appear in the Appalachian-Caledonian-Uralian belt, and Mesozoic (ca. 150 Ma) ophiolites dominate the Alpine-Himalayan belt (Fig. 20).

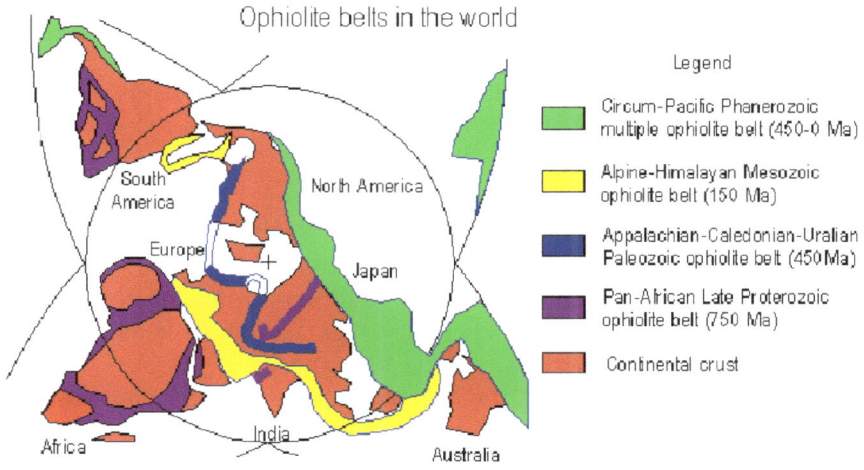

Figure 20. Distribution of ophiolitic belts.

It is interesting to note that within Guatemala we could have had the development of two pulses of ophiolites: older ones related to the Circum-Pacific Phanerozoic belt and younger ones from the Alpine-Himalayan Mesozoic one. This could explain the older age of the North and South Motagua ophiolitic belts.

Comparing Guatemalan Ophiolitic Belts

For this comparison, I will introduce three models of Alaskan ophiolitic belts that could be used as general models for the ophiolitic belts of Guatemala.

The North and South Motagua Ophiolitic Complex

It is my working theory, that the oldest ophiolitic complexes in Guatemala are represented by the North and South Motagua complexes within the Central Sector. They outcrop as narrow belts along the Motagua and Cabañas faults within the Motagua valley. Both ophiolitic belts consist of high pressure - low temperature (HP-LT) metamorphosed and serpentinized mantle harzburgites and foliated gabbros, followed by a thick basaltic pillow lava sequence (Fig. 21) showing mid-ocean ridge affinity of El Tambor Group (Beccaluva et al., 1995). They are both unconformable overlain by the Eocene continental flischoid polymictic sandstones and conglomerates of the Subinal Fm (Fig. 22).

Figure 21. Pillowed basalts from the south Motagua ophiolitic complex.

Figure 22. Flischoid polymictic sandstones and conglomerates of the Paleocene Subinal Fm. crops out between Kms. 146-151 of the CA 9 road in Guatemala.

The SM unit overthrusts the Paleozoic continental basement (Las Ovejas Group) of the Chortis block, while the NM unit overthrusts the Paleozoic metamorphic terrains (Chuacús Series) of the Sierra de Chuacús and Sierra de Las Minas.

The petrographic analysis of a group of samples from these belts shows that the most probable magmatic event was the partial melting process (Valls, unpublished).

Figure 23 shows a comparison between the Motagua ophiolitic complexes to the Union Bay ultramafic complex from Alaska. We can observe a general similarity in the zonal distribution of the more olivine rich rocks towards the center of the complex, surrounded in a metamorphic envelope of greenschists.

Figure 23. Comparing the Motagua complexes to the Union Bay ultramafic complex from Alaska.

The Juan de Paz – Los Mariscos Ophiolitic Complex

The Juan de Paz – Los Mariscos ophiolitic belt (JPZ) is composed of usually serpentinized lherzolites and websterites, layered gabbros, dolerites, basalt, and scarce andesites (Fig. 24).

Figure 24. Incipient boudinage on the lherzolites of the Juan de Paz - Los Mariscos ophiolitic complex.

It has been interpreted by Beccaluva et al. (1995) as island-arc magmatic sequences associated with sub-arc mantle rocks. This unit overthrusts the Paleozoic metamorphic basement (Chuacús Series) of the Sierra Las Minas and Montañas del Mico. The JPZ complex is usually covered by mafic volcanoclastics and andesitic breccias, passing upwards to carbonated breccias and calcarenites, with sandstone and microconglomerates containing felsic volcanic fragments of the Late Cretaceous Cerro Tipón Fm.

The JPZ complex is less metamorphosed than the Motagua complexes and shows more boudinage than the other complexes. It also shows effects of post-hydrothermal alteration, like the formation of botryoidal masses of magnesite.

The irregular pattern of the Rb:Ni ratio in samples from the JPZ ophiolitic complex indicates the possibility of later mixing event, which was confirmed by the La/Sr vs. 1/La ratio.

Figure 25 shows a transverse comparison between The Juan de Paz – Los Mariscos and Sierra de Santa Cruz ophiolitic complexes with the Goodnews Belt ultramafic complex in Alaska.

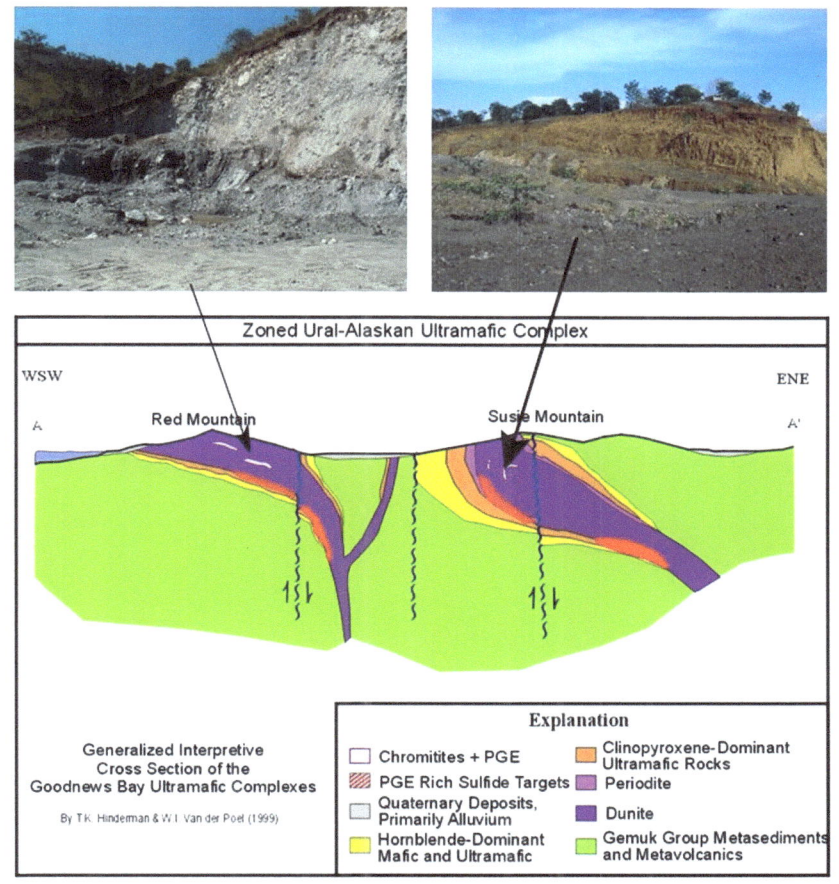

Figure 25. The cross section of the Goodnews Bay ultramafic complex compares well to a N-S section across the Lake Izabal and the Juan de Paz - Los Mariscos and the Sierra de Santa Cruz ophiolitic belts.

The Sierra de Santa Cruz Ophiolitic Complex

This complex appears to be the least metamorphosed in the Motagua sector. Petrologically, it is similar to the JPZ complex, being composed mainly of a juxtaposition of olivine-rich lherzolites and dunites with olivine-rich websterites and normal lherzolites with scarce cumulate gabbros and very few basaltic dykes or other volcanic rocks. All known Guatemalan lateritic deposits and most of the new targets are located within this complex.

The Sechol Unit

The Blashke Islands complex with its circular structures is very similar to the Sechol unit within the Sierra de Santa Cruz ophiolitic complex (Figs. 26 - 27).

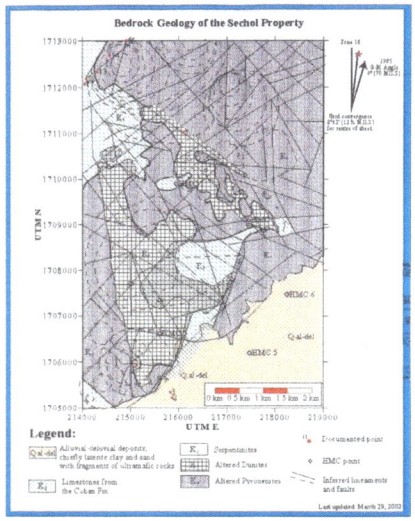
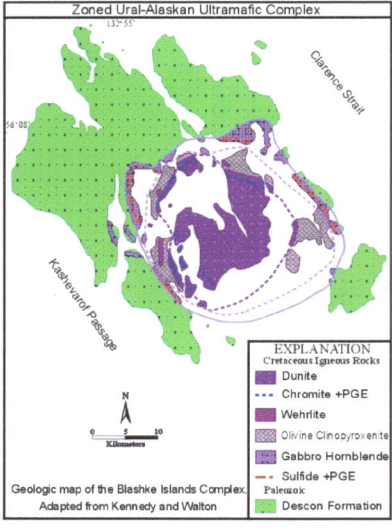

Figure 26. Circular structures at Sechol unit within the Sierra de Santa Cruz complex.

Figure 27. Circular structures at Blashke Islands complex in Alaska.

Once the general characteristics of the ophiolitic complexes have been introduced, I would like to present in more detail one of the most important geological processes in the formation of lateritic deposits - the serpentinization of these rocks.

Products of Serpentinization

Serpentinization requires large amounts of water ranging from 9% for a websterite, 15% for a lherzolite, and up to 19% for a dunite. The serpentinization is triggered by the percolation of sea water into the rocks and it also triggers an increase in the volume of the ultramafic rocks. The resulting stress could be an obduction/displacement mechanism within the ultramafic belt.

The final product of the serpentinization of an ultramafic intrusive depends on the petrology of the initial rock. Table 1 summarizes these products.

Table 1. Results of the serpentinization of ultramafic rocks.

Original petrology	Name of the rock	Water, %	Results of the serpentinization	Observations
Olivine (Forsterite)	Dunite	19	Serpentinite and brucite	Québec (Lac Chrysotile, See eq. 1)
Olivine + clinopyroxene	Peridotite (Lherzolite)	15	Serpentinite	Guatemala (See eq. 2)
Olivine + Ortho and clinopyroxene	Pyroxenite (Websterite)	9	Serpentinite and talc	Guatemala (See eq. 3)

Equation 1. Serpentinization of dunites (Dunites, Olivine-rich Lherzolites).

$3MgSiO_4 + 3H_2O \leftrightarrow Mg_2Si_2O_5(OH)_4 + Mg(OH)_2 + SiO_2 + O_2g\uparrow$

Equation 2. Serpentinization of peridotites (Lherzolites).

$MgSiO_4 + MgSiO_3 + 2H_2O \leftrightarrow Mg_2Si_2O_{25}(OH)_4$

Equation 3. Serpentinization of pyroxenites (Websterites).

$4MgSiO_3 + 3H_2O + 6SiO_2 + \frac{1}{2}O_2 \leftrightarrow Mg_2Si_2O_{25}(OH)_4 + 2MgSiO(OH)_2$

Table 2 and Fig. 28 show the systematic used in naming these ultramafic rocks.

Table 2. Systematic used in naming ultramafic rocks.

Field Name	Proper Name	Observation
"Dunite"	Dunite or Olivine-rich Lherzolite	Fig. 92
"Peridotite"	Lherzolite	Fig. 93
"Pyroxenite"	Olivine-rich Websterite	Fig. 94

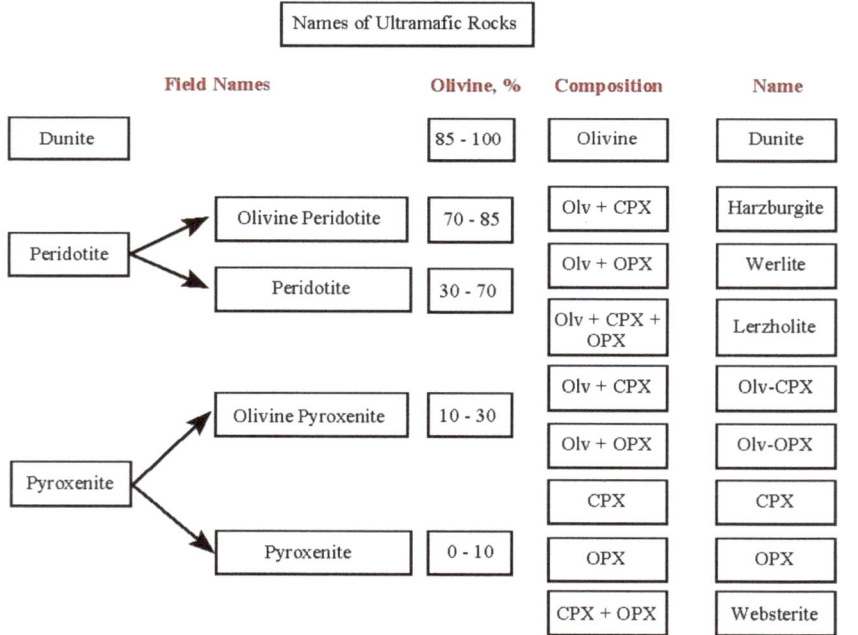

Figure 28. Systematics of the ultramafic rocks used during the exploration of ophiolitic belts in Guatemala.

Figure 29. Olivine-lherzolite (field name "dunite") showing the characteristic alteration rim that helps its easy identification.

Figure 30. Typical contact between an olivine-lherzolite (currently weathered into a saprolite) and a normal lherzolite (field name "peridotite"), with very limited lateritic development.

Figure 31. Olivine-websterite (field name "Pyroxenite"), with no presence of weathering.

Petrography of the Guatemalan Ophiolitic Belts

The following descriptions are based on a study completed by University of Toronto specialist Dr. Eva E. Schandl of several samples from the Guatemalan ophiolitic belts.

Olivine-rich Lherzolite

Sample number: 59068; UTM: 800164E   1682852N

Estimated mineralogy of the original rock prior to serpentinization: 70% olivine, 20% orthopyroxene, 10% clinopyroxene + chrome spinel.

The rock consists predominantly of olivine, lesser clinopyroxene, orthopyroxene, serpentine and some Cr-spinel (Fig. 100). Coarse grained. Olivine is fragmented and granulated and the fragmented grains are cross-cut by narrow serpentine veinlets. The pyroxenes (ortho and clino) are partly serpentinized and some contain olivine inclusions. Fine-grained magnetite that crystallized from the breakdown of olivine and the pyroxenes, is variably oxidized and occur as fine-grained narrow seams. Granulation of the rock is apparent from the presence of fine-grained, anhedral, granulated, and unserpentinized olivine aggregates which formed from the partial breakdown of the larger grains. Serpentine also occurs as massive veins that cross-cut the rock fabric, and contain minute, anhedral magnetite. The pyroxenes are partly or completely replaced by lizardite bastite, and olivine contains fracture-filling lizardite veinlets.

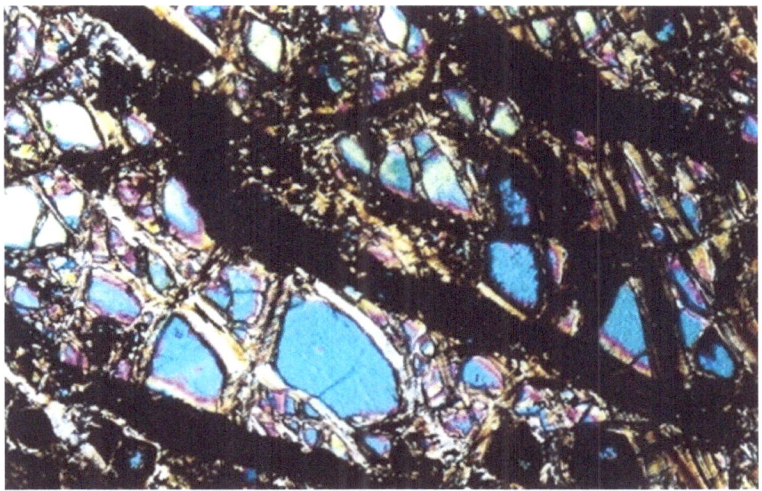

Figure 32. Serpentinized olivine with black (oxidized) magnetite veins. Width of the photo: 4 mm Crossed Nichols.

Lherzolite

Sample number: 59070; UTM: 800164E   1682852N

Estimated mineralogy of the original rock was: 55% olivine, 35% orthopyroxene, 10% clinopyroxene + chrome spinel.

Partially serpentinized, fragmented olivine, fractured and deformed orthopyroxene and clinopyroxene make up the rock. Coarse grained. Mesh texture lizardite (serpentine) is Fe-stained and serpentine veins that cross-cut the rock fabric contains fine-grained, partly oxidized magnetite seams in the center. Coarse-grained orthopyroxene with exsolution lamellae show evidence of extensive deformation, as the lamellae are strongly deformed, fractured, and the fractures are filled by fine-grained granular secondary clinopyroxene within the serpentine veins (Fig. 33).

Figure 33. Large deformed orthopyroxene (with exsolution lamellae) cut by serpentinite and small clinopyroxene veinlets. Width of the photo: 4 mm. Crossed Nichols.

Olivine-rich Websterite

Sample number: 59076; UTM: 798951E   1684986N

Estimated mineralogy of the original rock was: 35% olivine, 40% orthopyroxene, 25% clinopyroxene + chrome spinel.

The protolith is difficult to determine, due to the oxidation of serpentinized grains (which destroyed the diagnostic birefringence of the minerals) and the oxidation of serpentine. Relict orthopyroxenes are partly altered to bastite and mesh texture serpentine completely replaced what may have been originally olivine (Fig. 34). Orthopyroxene with narrow exsolution lamellae predominates over clinopyroxene in the rock. Two serpentine generations were identified; the first generation is the replacement mesh lizardite after olivine and bastite after the pyroxenes, and the second generation serpentine is represented by cross-cutting veinlets consisting of nickeliferous (up to 1.59% NiO) spherulitic serpentine. These veins are rimmed by Ni-free serpentine.

Figure 34. Bastite (yellow-orange) replaces orthopyroxene on this olivine-websterite. Width of photo: 4 mm. Crossed Nichols.

Hazburgite

Sample number: BVP Doc 1; UTM: 798689E   1684223N

Moderately serpentinized, coarse-grained harzburgite (Figs 35 - 38). The rock consists predominantly of olivine, orthopyroxene, and minor clinopyroxene. Anhedral and symplectic chromite are interstitial to the pyroxenes. Mesh lizardite rims the fragmented olivine and also occur as veins that cross-cut domains within the thin section. Some are serrate veins and form a network within olivine and pyroxene grains. Fe-staining of serpentine is common. Magnetite is relatively rare, suggesting that the rock is either Fe-poor or the magnetite that was produced during serpentinization, was later oxidized, locally giving the serpentine a rusty color. The rock shows evidence of partial recrystallization, as some olivine recrystallized to fine-grained aggregates.

Olivine and serpentine in the rock contain 0.4 and 0.6 wt% NiO, respectively. A few small pentlandite grains are interstitial to the serpentinized olivine. They are fragmented and the fractures are filled by hematite.

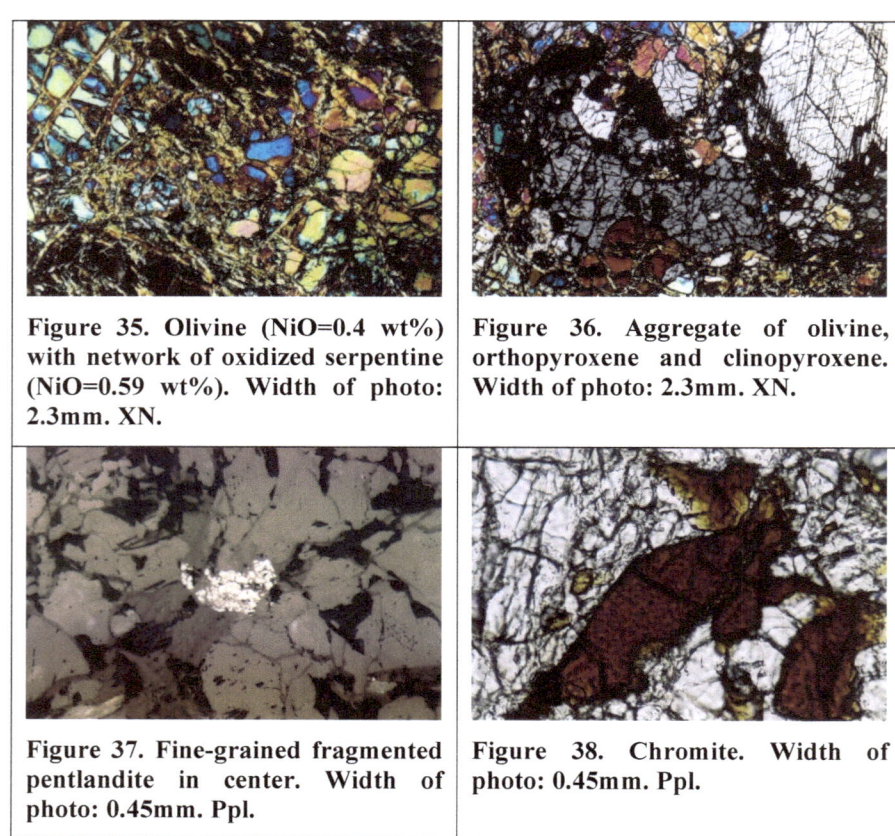

Figure 35. Olivine (NiO=0.4 wt%) with network of oxidized serpentine (NiO=0.59 wt%). Width of photo: 2.3mm. XN.

Figure 36. Aggregate of olivine, orthopyroxene and clinopyroxene. Width of photo: 2.3mm. XN.

Figure 37. Fine-grained fragmented pentlandite in center. Width of photo: 0.45mm. Ppl.

Figure 38. Chromite. Width of photo: 0.45mm. Ppl.

Table 3 shows the geochemical characteristics of some these samples.

Table 3. Geochemical characteristics of some of the petrological samples described above.

Sample	SiO_2	Al_2O_3	CaO	MgO	Na_2O	K_2O	Fe_2O_3
59068	39.61	1.07	0.85	40.03	0.009	0.04	7.97
59070	42.93	1.5	0.98	36.38	0.009	0.04	10.33
59076	41.91	1.5	0.63	34.62	0.009	0.04	7.86

Sample	MnO	TiO_2	P_2O_5	Cr_2O_3	L.O.I.	Sum
59068	0.12	0.04	0.009	0.4	9.7	99.77
59070	0.14	0.03	0.009	0.5	6.2	98.94
59076	0.09	0.04	0.009	0.43	11.5	98.54

Sample	Ni	Co	Cr	Cu	Zn	Nd	
59068	2200	120	2680	16	29	0.2	
59070	8180	133	3070	24	47	134	
59076	>10000	97.9	2880	16	35	160	

Sample	Pr	Rb	Sm	Tb	Tm	Y	Yb
59068	0.07	0.8	0.09	0.04	0.04	0.6	0.09
59070	37.3	0.7	19	3.27	1.47	148	8.5
59076	40.2	0.6	31.1	5.5	2.28	301	9.4

Sample	Cs	Dy	Er	Eu	Ga	Gd
59068	0.09	0.08	0.09	0.04	1	0.06
59070	0.6	17.5	12	4.45	2	21.3
59076	0.7	29.7	22.3	7.96	2	38.5

Sample	Ge	Hf	Ho	In	La	Lu	Zr
59068	0.9	0.9	0.04	0.1	0.3	0.04	3.2
59070	2	0.9	3.87	0.1	237	1.35	2.6
59076	1	0.9	7.29	0.1	193	1.57	2.3

Model for the Evolution of the Guatemalan Ophiolitic Belts

Phase I- Magmatic Chamber

The first stage of the evolution of these ophiolitic belts, took place in a hot and isolated magmatic chamber, where a low rate cooling combined with fractional crystallization created the original ultramafic rocks.

The crystallization followed the Bowen's series and the precipitation of minerals within the chamber followed the specific gravity of the formed minerals as shown in Figure 39.

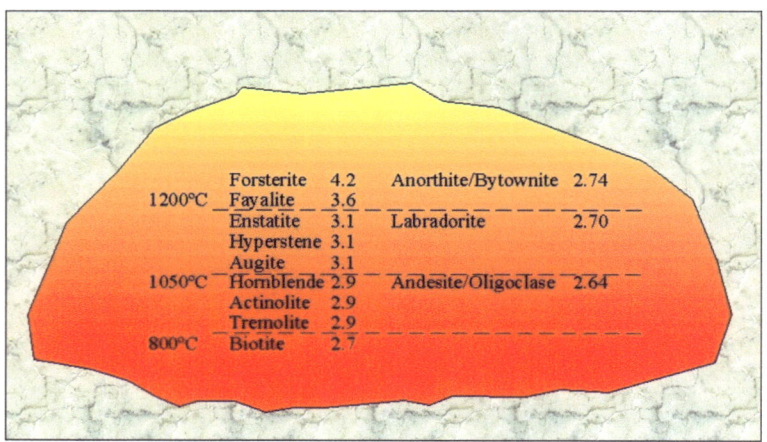

Figure 39. Crystallization and precipitation of olivine and pyroxenes as temperature in the magma chamber drops to 1050°C. Notice that plagioclase will tend to concentrate at the top of the chamber due to their lower density.

The fact that the described rocks only show the presence of olivine and pyroxenes indicates that the crystallization process stopped at a temperature over 1050°C. Their coarse grains also indicate a slow rate of cooling.

Phase II- Obduction

The second phase was the obduction of the ultramafic rocks. This obduction was related to the collision of the NW end of the Caribbean Plate with the Maya and the Chortis Continental Blocks. While there are differences in the age of obduction of different complexes along the Motagua and Polochic Faults, the general conditions of pressure and temperature after the obduction process should have been similar to all complexes (Fig. 40) and were very favorable for the start of the serpentinization process.

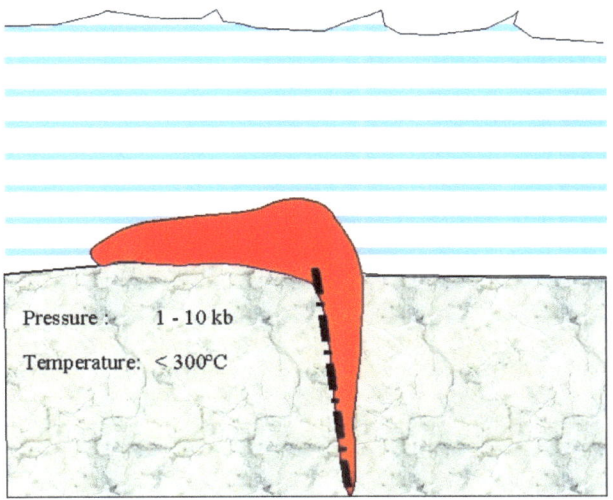

Figure 40. Obduction of the ultramafic rocks along the main sinistral faults of Central Guatemala.

In these conditions, both varieties of olivine started to react with the heated sea water as shown in equations 5 and 6.

Equation 5. Reaction of Fayalite with heated sea water during the obduction process.

$3Fe_2SiO_4 + 2H_2O \leftrightarrow 3SiO_2 + 2Fe_3O_4 + 2H_2$ (aq)

Equation 6. Reaction of Forsterite with heated sea water during the obduction process.

$Mg_2SiO_4 + H_2O \leftrightarrow SiO_2 + 2MgO + H2O$

Phase III- Serpentinization Process

The third phase is the serpentinization of the obducted rocks. The general equations for dunites, peridotites (lherzolites), and pyroxenites (websterites) were presented earlier (Equations 1 – 3). Following are more detailed equations of the reactions that took place under these conditions.

We will start first with the continuation of the reactions represented in equations 5 and 6 (see equations 7 and 8).

Equation 7. Formation of Fe-Chrysotile from Fayalite.

$3Fe_2SiO4 + 2H_2O \leftrightarrow 3SiO_2 + 2Fe_3O4 + 2H_2 \text{ (aq)}$

$2SiO_2 + Fe_3O_4 \leftrightarrow Fe_3Si_2O_5(OH)_4$

Equation 8. Formation of Serpentinite from Forsterite.

$Mg_2SiO_4 + H2O \leftrightarrow SiO_2 + 2MgO + H_2O$

$SiO_2 + 2MgO \leftrightarrow Mg_2Si_2O_5(OH)_4$

Equation 9. Formation of Garnerite from olivine.

$H_2 + SiO_2 + 2MgO \leftrightarrow H_2Mg_2SiO_4$

Equation 10. Formation of Bastite from orthopyroxene.

$(Mg, Fe)SiO_3 + H_2O \leftrightarrow SiO_2 + Mg_2Si_2O_6$

Equation 11. Formation of Brucite from Clinopyroxene.

$Ca(Mg, Fe)Si_2O_6 + H_2O \leftrightarrow SiO_2 + (Mg, F)(OH)_2$

The fact that none of these reactions continued to their completion, e.g. we still have plenty of magnetite and even sulphides and olivine relicts in the weathered rocks, is an indication that the serpentinization process was interrupted (end of heat source? sea regression?).

Of special interest is the production of $H_2(g)$ in equation 7. Gibbs and Hui (1971) stipulated that the H_2 could react with effusive CO in the presence of nickel as a catalyzer to form hydrocarbons.

Phase IV- Oxidation process

Oxidation represents the fourth phase of this process (see eq. 12 and 13). Weathering affected more strongly the olivine rich rocks within these belts, where full laterite profiles were formed (Geosol Izabal).

Equation 12. Formation of goethite from magnetite.

$Fe_3O_4 \leftrightarrow Fe_2O_3 + FeO$

$H_2O + CO_2 \leftrightarrow H_2CO_3$

$FeO + H_2CO_3 \leftrightarrow FeCO_3 + H_2O$

$Fe_2O_3 + 3H_2CO_3 \leftrightarrow Fe(CO)_3 + 3H_2O$

$Fe_2(CO_3)_3 + 4H_2O \leftrightarrow Fe(OH)_3 + 3CO_2 + H_2O$

$Fe(OH)_3 \leftrightarrow FeO(OH) + H_2O$

$Fe_2(CO_3)_3 + 4H_2O \leftrightarrow 2FeO(OH) + 3CO_2 + 3H_2O$

Equation 13. Oxidation of sulphides to form goethite.

$2NiS + 7O_2 + 2H_2O \leftrightarrow 2NiSO_4 + 2H_2SO_4$

$4NiSO_4 + O_2 + 2H_2SO_4 \leftrightarrow 2Ni(SO_4)_3 + 2H_2O$

$Ni(SO_4)_3 + 6H_2O \leftrightarrow 2Ni(OH)_3 + 3H_2SO_4$

Unique Characteristics of the Geosol Izabal

When comparing the laterites from the SSC complex with other wet and dry laterites of the world, the uniqueness of this deposit becomes clear (Fig. 41).

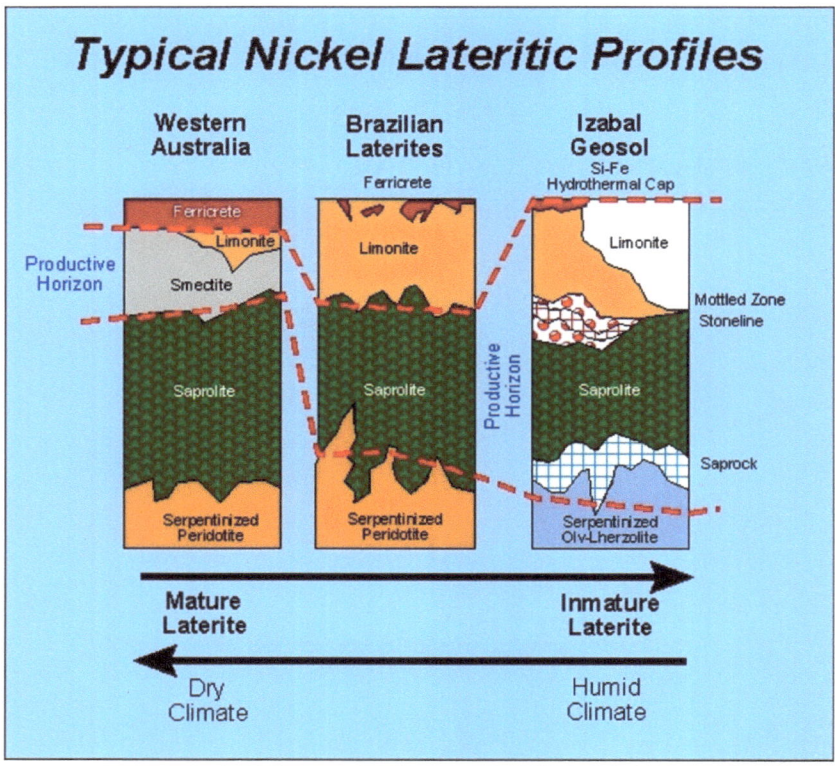

Figure 41. Comparison of the Izabal Geosol with other dry and wet laterites of the world.

The most important differences here are the age of intrusion and the degree of metamorphism of the SSC unit. As shown previously in Figure 82, the ages of the ultramafic complexes seem to grow older in a west – east direction. The North and South Motagua ophiolitic complexes are representative of zones of HP-LT conditions, while the degree of metamorphism at the Juan de Paz – Los Mariscos is more limited to zones of intense mylonitization. According to my mapping, the intrusion of the SSC unit occurred during the Early Tertiary and here the metamorphism was limited to the selective serpentinization of the complex.

Another indication of the immaturity of these laterites is the presence of up to 30% of magnetite as an average for the whole profile of the Geosol, as well as the absence of the pisolitic iron crust.

We completed a study of the magnetic fractions of samples from different belts in Guatemala at Inotel Lab, in Sherbrook, with Dr. Jean-Marc Lalancette (Figures 42 - 49).

Figure 42. Dr. Jean-Marc Lalancette mixes a bag with pulverized material before taking a sample.

Figure 43. An initial sample of 75 grams was taken from each rock.

Figure 44. Mixing the sample with water, prior to the magnetic separation.

Figure 45. Separation of the magnetic fraction with a magnet and water.

Figure 47. Magnetic (left) and non-magnetic fractions after 5 hours of drying at 910C.

Figure 46. Filtering the non-magnetic fraction in a vacuum filter.

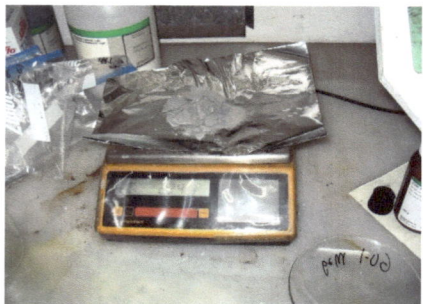

Figure 48. Weighing the magnetic fraction.

Figure 49. Weighing the non-magnetic fraction.

Table 4 shows the results of this study and also includes historical data from the Tailings from Quebec.

Table 4. Results of the analysis of the magnetic composition of samples from Guatemala and Quebec.

Lithology	Location	Total weight, g	NMgt, g	Mgt, g	Mgt, %
Lherzolite	Juan de Paz	75.0	50.1	22.4	30.0%
Olivine lherzolite	Juan de Paz	75.0	61.0	12.6	17.0%
Websterite	Juan de Paz	75.0	58.3	15.1	20.0%
Boudine	Sierra de Santa Cruz	75.0	66.3	6.1	8.0%
Websterite	Sierra de Santa Cruz	75.0	49.7	18.7	25.0%
Olivine lherzolite	Sierra de Santa Cruz	50.0	34.4	11.2	22.0%
Harzburgite	North Motagua	75.0	24.5	48.9	65.0%
Limonite	Sierra de Santa Cruz	250.0	87.5	162.5	65.0%
Saprolite	Sierra de Santa Cruz	278.0	191.8	86.2	31.0%
Tailings	Lac Chrysotile				35.0%
Tailings	J-M				28.0%
Tailings	Carey				43.0%
Tailings	Maternal				29.0%
Tailings	BC-1				19.0%
Tailings	BC-2				28.0%
Tailings	Beaver				25.0%
Tailings	Nomandie				19.0%
Tailings	Bell				25.0%
Tailings	Average value				28.0%

Another interesting difference is the grain size of these saprolites. As shown in Table 5, even at -60 Mesh there is abundant still sandy material in these saprolites.

Table 5. Granulometric analysis of saprolites from the Geosol Izabal.

Sample	+10 Mesh	+40 Mesh	+60 Mesh	-60 Mesh	Total, g
74755	767.00	333.00	116.00	195.00	1411.00
74753	610.00	402.00	104.00	169.00	1285.00
73807	501.00	494.00	159.00	162.00	1316.00
73801	336.00	257.00	83.00	164.00	840.00

Sample	+10 Mesh	+40 Mesh	+60 Mesh	-60 Mesh
74755	54.00%	24.00%	8.00%	14.00%
74753	47.00%	31.00%	8.00%	13.00%
73807	38.00%	38.00%	12.00%	12.00%
73801	40.00%	31.00%	10.00%	20.00%

	+10 Mesh	+40 Mesh	+60 Mesh	-60 Mesh
Average	44.97%	30.75%	9.57%	14.70%
St. deviation	7.45%	5.70%	1.86%	3.27%
Variability	16.57%	18.54%	19.46%	22.27%
Maximum	54.36%	37.54%	12.08%	19.52%
Minimum	38.07%	23.60%	8.09%	12.31%

Finally, the fact that most of these laterites develop mainly over dunites and olivine-rich lherzolitic rocks, with very limited to non-existent development over lherzolites and olivine-rich websterites, is also an indication of the young age of this deposit.

References

Beccaluva L., S. Bellia, M. Coltorti, G. Dengo, G. Giunta, J. Mendez, J. Romero, S. Rotolo, and F. Siena, 1995. The northwestern border of the Caribbean Plate in Guatemala: new geological and petrological data on the Motagua Ophiolitic belt. Ofioliti, 20 (I):1-15.

Giunta G., L. Beccaluva, M. Coltorti, D. Cutrupia, B. Mota, E. Padoa, F. Siena, C. Dengo, G. E. Harlow, and J. Rosenfield, 2002. The Motagua Suture Zone in Guatemala, Field Trip – Guide Book, publication supported by Ofioliti Int. Journal (M. Marroni and L. Pandolfi eds.).

Trusova I. F. and V. I. Chernov, 1982. Petrografia magmaticheskij I metamorficheskij gornij parod (in Russian).Nedra, Moscow, 272 pp.

Valls Alvarez, R. A., 2002. Summary Report on the Geology and Mineral Resources of the Sechol Nickel-Magnesium Laterite Deposit. Guatemala, Central America. SEDAR, 332 pp.

Valls Alvarez, R. A., 2003. The Geosol Izabal – a different type of nickel-cobalt laterite in Central America, Internal Report for Jaguar Nickel Inc.

Valls Alvarez, R. A., (unpublished). Field Petrology. A Practical Approach to the Study of Petrologic Data. 350 pp.

Notes

About the Author

As a professional geologist with thirty-two years in the mining industry, I have extensive geological, geochemical, and mining experience, managerial skills, and a solid background in research techniques, and training of technical personnel. I am fluent in English, French, Spanish, and Russian. I have been involved in various projects world-wide (Canada, Africa, Russia, Indonesia, the Caribbean and Central and South America). Projects included from regional reconnaissance to local mapping, diamond drilling and RC-drilling programs, open pit and underground mapping and sampling, geochemical sampling and interpretation, and several exploration techniques pertaining to the search for diamonds, PGM, gold, nickel, silver, base metals, industrial minerals, oil & gas, and other magmatic, hydrothermal, porphyritic, VMS and SEDEX ore deposits. Special strengths are related to acquisition of new properties, geochemical and geological studies, management and organization, geomathematical analysis and modelling, compositional data analysis, structural studies, database design, QA&QC studies, exploration studies and writing technical reports. P.Geo. registered in the province of Ontario.

www.ingramcontent.com/pod-product-compliance
Lightning Source LLC
Chambersburg PA
CBHW040926180526
45159CB00002BA/633